GRAYISMS

GRAYISMS

AND OTHER THOUGHTS ON LEADERSHIP
FROM GENERAL AL GRAY, USMC (RETIRED)
29TH COMMANDANT OF THE MARINE CORPS

Compiled by
PAUL OTTE

 POTOMAC INSTITUTE PRESS

Copyright © 2015
Potomac Institute for Policy Studies
&
Paul Otte,
All rights reserved.

ISBN: 978-0-9898556-8-6

Our sincere thanks to the USMC for granting
permission to use the portraits on the covers.

The author has made every effort to trace copyright holders and to obtain their
permission for use of this material. The author and publisher apologize for any
omissions and would be grateful if notified of any corrections that should be
incorporated in future reprints or editions of this book. Government-produced
materials are not copyright protected (17 U.S.C. § 105). The author and publisher
do not hold the copyright to works of the United States government appearing
in this publication. The author and publisher cannot assume responsibility for the
validity of all materials or the consequences of their use; the view and opinions
expressed do not necessarily reflect those of the Institute and publisher.

POTOMAC INSTITUTE FOR POLICY STUDIES
POTOMAC INSTITUTE PRESS
901 N. Stuart St, Suite 1200, Arlington, VA, 22203
www.potomacinstitute.org
Telephone: 703.525.0770; Fax: 703.525.0299
Email: webmaster@potomacinstitute.org

"I don't run a democracy. I train troops to defend democracy and I happen to be their surrogate father and mother as well as their commanding general."

> *Major General*
> *Alfred M. Gray, USMC*
> *CG 2nd Marine Division and*
> *the Carolina Marine Air Ground*
> *Task Force (MAGTF)*

Table of Contents

Introduction 1
About the Potomac Institute for Policy Studies . . 3
About Grayisms 7
Grayisms 9
Additional Grayisms 53
Additional Stories. 63
General Al Gray Biography 71
To U.S. Marines. 77
Additional Information 79
Works Cited 81

Introduction

From the very beginning of working with General Gray, there were several things that became very obvious. Foremost, are how much he loves his Marines along with the Sailors who serve with them, and how much they love and respect him in return. He has great admiration for all our Armed Forces. Another thing is how humble he really is. It's never been, and never will be about him. And part of that respect is how often people have General Gray stories to share with you – we gave them a name – Grayisms.

Lastly, we came to realize how little many Marines really know about him, especially in his earlier years when (it may be hard for many to believe) his impact on the Corps may have been even more significant than his years as the 29th Commandant.

That's because he wasn't willing to sit still long enough to capture his history. We finally accomplished that when he sat for a series of video taped interviews in preparation for a series of books being written by the historian (and former Marine Major), Scott Laidig.

Always more willing to talk about others than himself, we (Scott, Master Sergeant J.D. Baker [USMC Retired], Lieutenant Colonel Eric Carlson [USMC Retired], and me) sat for over forty hours as General Gray spoke without notes, but with powerful emotions about the Corps and the Marines he continues to serve even today.

We were able to gain greater insight into this very special Marine who "took what he got, and made what he wanted." This book is a compilation of the many sayings we have heard and heard repeated, as they have been shared from one Marine to another.

And thanks to Brigadier General Dave Reist, USMC (Retired) for the idea, we added just a few of the stories told by others about General Gray and the impact he had on them.

But the stories do not stop here. We have captured only a small part of General Gray. Ask any Marine that served with him, or have heard their own stories about the General, and odds are they will have a story, a *Grayism,* to share with you.

Paul Otte
Corporal, USMC (1961-1965)
November 10, 2014

About the Potomac Institute for Policy Studies

The Potomac Institute for Policy Studies is an independent, 501(c)(3), not-for-profit public policy research institute. The Institute identifies and aggressively shepherds discussion on key science, technology, and national security issues facing our society. The Institute remains fiercely objective, owning no special allegiance to any single political party or private concern. With over nearly two decades of work on science and technology policy issues, the Potomac Institute has remained a leader in providing meaningful policy options for science and technology, national security, defense initiatives, and S&T forecasting. The Institute hosts academic centers to study related policy issues through research, discussions, and forums. From these discussions and forums, we develop meaningful policy options and ensure their implementation at the intersection of business and government.

These Centers include:

- Center for Revolutionary Scientific Thought, focusing on S&T futures forecasting;

- Center for Adaptation and Innovation, chaired by General Al Gray, focusing on military strategy and concept development;

- Center for Neurotechnology Studies, focusing on S&T policy related to emerging neurotechnologies;

- Center for Regulatory Science and Engineering, a resource center for regulatory policy; and

- International Center for Terrorism Studies, an internationally recognized center of expertise in the study of terrorism led by Professor Yonah Alexander.

The Potomac Institute's mission as a not-for-profit is to serve the public interest by addressing new areas in science and technology and national security policy. These centers lead discussions and develop new thinking in these areas. From this work the Potomac Institute develops policy and strategy for their government customers in national security. A core principle of the Institute is to be a "Think and Do Tank" rather than just conduct studies that will sit on the shelf, the Institute is committed to implementing solutions.

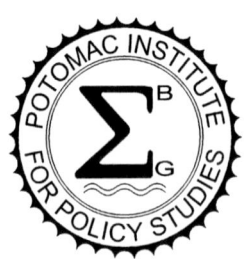

It seems that everyone is aware of General Gray's great successes. However, I have witnessed, first hand, his failure – at retirement. As Potomac Institute for Policy Studies Senior Fellow; Chairman, Board of Regents; and member, Board of Directors, General Gray sets an example that others find hard to follow. He can leave others (many half his age) in the dust.

General Gray continues to serve his country in thousands of behind-the-scenes ways. Al Gray – the mentor, teacher, friend, and surrogate father – continues to impact the lives of everyone he encounters.

I am proud to be one of the lucky 'adequate' ones ('adequate' is high praise from General Gray).

Mike Swetnam
Chairman of the Board/CEO
Potomac Institute for Policy Studies

About Grayisms

What are Grayisms?

Grayisms are recurring statements made by General Alfred M. Gray, Jr. USMC (Retired) who served as the 29th Commandant of the Marine Corps, a member of the Joint Chiefs of Staff, and as such, before his retirement in 1991 a military advisor to Presidents Ronald Reagan and George H.W. Bush. Many believe General Gray transformed the Marine Corps during his years of service with his unique form of leadership. He helped prepare the Marine Corps for the 21st Century.

What follows is a compilation of statements made by General Gray during his 41 years in the Marines and in subsequent years. They are based on interviews with General Gray and recollections from other Marines and former Marines who have heard and adhered to the guidance found in the recurring statements remembered as "Grayisms".

Grayisms are the embodiment of simplicity (capturing the essence; making concepts easy to understand).

The listing is in no particular order of importance. Grayisms, like other words of guidance, become important when they are applied. As evidenced from the many times these statements were repeated, they are important to the Marines that remember them. If applied, they will become important to any leader.

Nor is the listing complete. There are many more Grayisms. Ask other Marines currently serving, retired, or former and you are likely to hear another story based on a Grayism, either from their personal experiences with General Gray, or passed on from one Marine to another. This is just a sampling.

Grayisms

Because General Gray began his career as an enlisted Marine, and continues to remain close to all "his" Marines, he will always remain "the enlisted Marines' Commandant." It only seems fitting that we turn to David W. Sommers, Sergeant Major, USMC (retired), selected by General Gray to be the Sergeant Major of the Marine Corps (the top enlisted Marine) to introduce our first Grayism...

What do you want it to look like when we are finished?

> General Gray embodies leading by example both in his words and his bearing. His demeanor is that of a Company Gunny. It is why enlisted Marines see him as the modern day Chesty Puller. He has never lost his '2nd Lieutenant' mentality when it comes to caring for 'his Marines.'
>
> He is as honest, sincere, and genuinely concerned as a 2nd Lieutenant on the day he loses his first Marine. It is his strong point.

> I was with him in the 2nd Division during the Beirut bombing. He took it very personally.
>
> Afterwards we went to almost every funeral (the ones we missed were occurring simultaneously). It took us awhile to get our feet back under us after Beirut, but he pulled us back together again and we learned from it. General Gray vowed to never again allow Marines to be placed in harm's way without a purpose, a mission, and an intent clearly stated.
>
> *Sergeant Major David W. Sommers*
> *USMC (retired)*

General Gray's common response to a request for action became: *"What do you want it to look like when we are finished?"*

This simple statement emphasizes the importance of having an expected, and stated, result before committing any resources. It is a fundamental part of any strategy.

All leadership requires committing people and resources to an end state and assuring the results are worth the resources. General Gray tells us "if you want them to follow, tell them the end state." None know this better than the leaders in our military whose resources include placing human lives in harm's way.

> From the envisioned end state, we can develop the operational objectives which, taken in combination, will achieve those conditions.
>
> *"Campaigning" Fleet Marine*
> *Force Manual (FMFM) 1-1*

You are the one responsible

General Gray has always seen himself as the one ultimately responsible for what happened to his Marines. In 1990, after a helicopter crash in Korea, Marines were medically evacuated to the Burn Center at San Antonio, Texas. Sergeant Major Sommers was with him when General Gray walked into the waiting room, told the families assembled there that he was responsible for their sons' injuries, and asked them what he could do to help. This took moral courage.

His actions symbolize the unspoken bond between Marines, a spirit that has led them through hard times. Even today, he is asked to many reunions, numerous special occasions, and despite a very busy schedule, he joins his Marines whenever he can – Why? It is a part of his sense of responsibility.

Marines have, or at least believe they have, a relationship with the Commandant. And the Commandant readily accepts that he's responsible for all they do, or fail to do.

It goes back to General John A. Lejeune, our 13th Commandant, when he said the relationship between officers and enlisted Marines is not one of a superior to a subordinate, but more like a teacher and a scholar; and he implored that all Marines should be responsible for their activities, and that you owe it to your Marines to see to it that each one is stronger morally, mentally, and physically when he leaves you than when he joins you.

And that's a cardinal thought process that Marine Commandants share and accept as a responsibility.

Anybody have any questions?

General Gray went to his Marines, talked to them, listened to them, and responded to their needs. He would always ask, "Anybody have any questions?" Sergeant Major Sommers remembered one Marine, a Staff Sergeant, who said he was being told he was too fat to reenlist, but he lifted weights and was, in his opinion, within the "percentage of fat" requirement.

General Gray told him to report to the Sergeant Major in the hall after the meeting. As you can imagine a crowd formed waiting to see what he would do. General Gray asked the Staff Sergeant if he had played football? The response was "yes, sir." The General then asked him to assume the three-point football stance and said if he could hold firm, he could stay in the Marines. If not, he would be out.

The Staff Sergeant took the General's blow, remained on his feet and in the Corps.

General Gray, as he did on many other occasions, said "Sergeant Major, take care of it."

> This kind of easy, empathetic relationship between the two men (General Gray and Sergeant Major Sommers) was evident at many dinner gatherings and other meetings, where Gray, when he had finished his speech, would call out, 'Sergeant Major, stand up!' And, pointing to Sommers, he would proclaim to the room, 'That's the guy who really runs the Marine Corps!'
>
> *"Uncommon Men, The Sergeant Majors of the Marine Corps" (by John C. Chapin, pg. 290)*

Take care of it

Sergeant Major Sommers took care of a lot of things. General Gray gave him mission guidance and trusted him to get it done. Sergeant Major Sommers was "in the General's hip pocket" for four years and never saw him put on a false face.

It was a challenge keeping up with General Gray. Sergeant Major Sommers said he was the hardest person he ever worked for. He demands a lot from others and himself. He was always on the move practicing what he preached – maneuver.

They went to the field frequently, often without advance notice. He had cards printed:

Do you know where your Commandant is today?

He dropped them on people's desk (often at recruiting stations) to let them know he had been there and was already gone.

> After every trip we got back together, and sometimes after every leg of a trip where we were to hit different bases. We would talk about areas that I found and that he found, and we would compare notes. General Gray was very good when I would tell him things, and would indicate to him that I'd like to take care of that myself. In turn, I would tell him if I knew something he should be aware of.
>
> *Sergeant Major Sommers*
> *In "Uncommon Men" (pg. 288)*

Hold them together and point them in the right direction

During *Desert Shield* and *Desert Storm* General Gray went to visit the troops often. The mission each time was to talk to as many young officers and NCOs as possible.

The message – the most critical leadership point is in the first 15 seconds. Hold your Marines together and point them in the right direction.

The hardest thing was when he had to come back to Washington. Sergeant Major Sommers said the General was worried about his Marines and the chemical threat.

> General Gray was very, very concerned with the possibility of heavy casualties. As a result, we spent a lot of time going from one unit to another . . . I sat in meetings with him continually. We would come out of the meetings and we would discuss what went on in the meeting, and it was always 'What do you think Sergeant Major?'
>
> I also often shared with him my worry for him personally. He was so concerned that I was becoming increasingly concerned about his own health . . . the man didn't sleep, he didn't eat. He was constantly traveling, constantly working, constantly preaching the message of preparedness.

Sergeant Major Sommers
In "Uncommon Men" (pg. 300)

The greatest tribute to General Gray's philosophy for leading came during *Desert Storm*. There wasn't an officer or an enlisted leader on the battlefield that was afraid to make a decision following the maneuver thought process, mission guidance, and commander's intent.

> We had focused so hard on our capability, on training, on maneuver warfare, on all the things we did in the liberation of Kuwait. To me it was the final examination. Does it work? Yes, it did . . . our Marines went through that place like a knife through butter.
>
> You saw young officers and young corporals and sergeants who, in the true sense of maneuver warfare . . . didn't have to worry about getting authority from a higher command . . . . They understood the commander's intent, and that freed them to do what they had to do.
>
> *Sergeant Major Sommers*
> *In "Uncommon Men" (pg. 300)*

General Gray put the Warrior back into the Marines' mentality and he did it with the tenacity of a bulldog. He is a tremendous visionary.

His concerns are for the 21st Century Marine. When the Marines came back from the desert, they put their equipment away and immediately began training for the next battle.

Don't paint rocks

General Gray didn't like Marines spending valuable time and effort preparing for his visits. He would say: *Don't paint rocks.*

Having served as a General officer for 15 years, including four as the Commandant of the Marine Corps, he was surely exposed to all forms of military ceremonies and the pomp that tends to go with them.

His point – don't waste your effort doing unnecessary things. Well-meaning people can go to great lengths doing things that are not required. Painting rocks in advance of a General's visit is just one example.

We all know organizations that "paint rocks" and leaders that knowingly allow it to occur (some might even expect it). Some "paint rocks" in more subtle but still wasteful ways.

The best leaders neither openly, nor tacitly, condone the painting of rocks. Instead they concentrate their resources on the things that matter.

> General Gray did not like big groups meeting his plane when he arrived at locations. His plane had no special markings – it was all gray.

> The General hated it when folks would be waiting on the tarmac. On one occasion, we landed at 29 Palms and as we taxied in, General Gray saw all the Generals lined up to greet him and he got upset. He had

put the word out for them to continue operations and not come to meet him. He did not like Marines standing around putting on this kind of dog and pony show.

At first, he did not get off the plane (he was talking on his briefcase phone to Washington), and he called the SMMC (Sergeant Major of the Marine Corps) to the front of the plane. He told the SMMC to get off first. The SMMC got off the plane and when the Generals saw him enter the hatch they thought it was General Gray, they all snapped to attention and saluted him as he came down the ramp. The SMMC returned the salute. They got the message.

Later, General Gray told everyone that the SMMC got in his staff car and drove away so he had to take the jeep – the Marines always got a good laugh out of it.

Lieutenant General George Flynn
USMC (Retired)

Know yourself, know your people, know your profession

This is General Gray's fundamental tenet of leadership. He will quickly point out he is not the first leader who believed in it. To which we would respond – he personified it.

To be a leader, one must begin with knowing oneself. The real you, not what you think you are, but what others know you are, and it includes knowing what you don't know.

Knowing your people goes beyond their name, rank, and serial number. It means knowing them as people, starting with each individual.

It all comes from the quality of people. One level of capability in which we are all equal is in our ability to love and care for people. Your family and the people you love are the #1 things in your life. The Marine Corps is the #1 thing you do.

General Gray was known for conducting inspections from behind. While the troops were standing in formation, he along with their Commanding Officer would stand behind them. Then General Gray would ask the CO to tell him about a person he pointed to – name, family, children, years in the Corps…

Many think no one may know their profession better than General Gray. His office, even today, resembles a small library. When deployed, he often took a "foot-locker" full of books to be read. And when he returned they were not only read, they were also marked up with highlights and notes.

And their contents were committed to memory. To this day, he will stop in conversation to get a book, to reinforce the point he is making (and it most likely will have been read many years before).

> In my humble opinion, the military is going overboard on information and underboard, if that's a word, on knowledge. And we ought to concentrate more on knowledge and what that means, and less on information.
>
> *General Al Gray*

> A leader without either interest in or knowledge of the history and theory – the intellectual content of his profession – is a leader in appearance only.
>
> *FMFM 1 "Warfighting"*
> *USMC Doctrinal Philosophy*

Take what you get, make what you want

This is one of the most often repeated phrases of General Gray. He uses it in many ways – in talking about his assignments, the Marines he has had the responsibility to lead, and in overcoming the obstacles he faced.

While others sought out assignments that would further their careers, General Gray accepted every duty given to him and made the best of it. In the long run, it was a key ingredient to his success as a professional in the Nation's Corps of Marines.

For those Marines General Gray was privileged to lead all through the years, he believed it was his responsibility to help them become better Warriors and better citizens in life. In all command assignments, he was dedicated to the simple thought that his Marines would be better people and part of a better unit, when he left them than when he joined them.

He was not impressed with commanders who always sought to get the very best people in their organizations at the expense of others who also had potential if someone helped them.

As a matter of fact, some of our greatest research, development, and acquisition programs in the later part of the 20th Century were led by Marines who were *not* considered to be in the highest categories for advancement purposes.

You cannot write a recipe for all situations

"Doctrine is a guide," General Gray tells us. He wrote in the Foreword to *Warfighting*:

> This book describes my philosophy on warfighting. It is the Marine Corps' doctrine and, as such, it provides the authoritative basis for how we fight and how we prepare to fight . . . . It is not intended as a reference manual, but is designed to be read from cover to cover . . . . this book does not contain specific techniques and procedures for conduct. It requires judgment in application.
>
> *General Al Gray*
> *Foreword to FMFM 1 "Warfighting"*

FMFM 1 *Warfighting* was written under General Gray's personal guidance when he was the 29th Commandant. It is a small book (less than a hundred pages). The thoughts contained in it are not just guidance, but a way of thinking in general – a philosophy.

Anything else defeats the whole purpose of having people who can think.

> FMFM 1 is useful in battle, in business, and in life. It's a methodology for everything you do every day. It's a way of life – a lifetime thought process.
>
> *General Al Gray*

It doesn't cost any money to think

Thinking can be our greatest resource. Thinking can be the source of new ideas, new approaches, and new solutions. Thinking doesn't cost any money (and it might just save money).

> We had a lot of time to think. Many assignments were on the lonely side, and traveling here and there so you had time to think and be curious about things. I often think one of the prime attributes that I can ascribe to upcoming officers and noncommissioned officers is have a healthy curiosity. I think that's very important.
>
> Somehow you have to provide a way for commanders and key staff officers at every level to have enough time to think. And that takes a lot of discipline.
>
> In times of limited fiscal resources (like the '70s), when you don't have the money, you can still think. We spent a decade studying our requirements and overall needs. And thus, when increased budgets became available in the 1980s (President Reagan buildup), we struck like Genghis Kahn. Some of the greatest innovative concepts were formed when money wasn't available.
>
> *General Al Gray*

And tied to this idea is the next thought, which many people have had...

Want a new idea – read an old book

An avid reader, General Gray established the Commandant's Reading List in 1988. Every Marine, from Private to General, is encouraged to read at least two books annually from a listing of recommendations (by rank).

Many might find it unique that Marines of all ranks are expected to read, but it continues to be a key part of their Professional Military Education. How many others could benefit by reading more about their profession?

Why an old book? Because the best ideas often come from making connections, and linking present problems with answers from the past can be a great source of new ideas.

> I have always been a reader and, of course, I spent so much time aboard ship. For example, I would take a footlocker full of books with me to the Mediterranean and the Mideast and read about the countries in that region and the impact of their religious history and culture. It's important to look at people through their eyes, not just our eyes. You name it, I read it.
>
> *General Al Gray*

First, care about what you are

General Gray tells us:

> Care first about what you are – your values, beliefs, and actions. Then you will care about how you look (and for the right reasons).
>
> Others follow you for what you are, because they believe in you and what you do. You look in a mirror to see how you look. You look in the faces of others to know what you are.
>
> *General Al Gray*

When we think about it, this Grayism captures the essence of leadership development programs. Many leadership development efforts focus on "how you look." Knowing the popular leadership writings, "talking the talk," identifying your preferred "leadership style." It is an "Outside In" approach.

Care first about what you are – your values, beliefs, and the way you think. Starting with your definition of leadership through your leadership philosophy, doing the right things, for the right reasons, "walking your talk," all based on the needs of your followers. It is an "Inside Out" approach.

Or, as General Gray would say, "it's really a matter of having things in the right order."

> The one Grayism that captures his style and sense of focus that I recall would be during the period he was a Battalion and Regimental Commander at Camp Lejeune. At that time, there was a general discussion about what should be included in a Marine's sea bag.
>
> There were a number of senior officers advocating that each Marine should be issued a set of Dress Blues and they should be worn on liberty and travel. HQMC sent out a message asking for input. The question was brought up in a large gathering and Gray was asked his opinion.
>
> His response was; 'we could be the best dressed outfit ever to get run off the battlefield.'
>
> *General John Sheehan*
> *USMC (Retired)*

You must care more about the people you are privileged to lead than about yourself

General Gray believes "Everything that gets done, gets done through people." Leaders must truly care more about the people they lead than themselves. To him, these are more than just words, they represent a way of life, his life, and he proved it by the example he set.

The examples we set may never approach the level of personal sacrifices made by General Gray, but we can put our people ahead of ourselves in many meaningful ways. Those who are unwilling to do so, are not leading and we doubt there are many that are willingly following them.

> Well, that's just the way we were – you try to do these things through setting the example and the way you are. If you believe in them and you believe in what you're doing, then that carries over to others. But it's just the way you are.
>
> I've always believed that to be an officer of Marines, for example, you have to be yourself and you should act like that. And when you try to act like somebody else, that's when you begin to have problems. And that's true for anybody, whether you're a corporal or a gunnery sergeant or a major or a general.
>
> *General Al Gray*

Sergeant Major Sommers describes his "succinct" original marching orders from General Gray:

> We'll go down the trail together on this thing, but I want to tell you one thing we are going to do. We are going to get around and see our Marines; we're going to take care of them; and we're going to have a good time doing it.
>
> *Sergeant Major Sommers*
> *In "Uncommon Men" (pg. 283)*

Tom Ricks says this about General Gray:

> Gray's effect on the Corps was enormous. Speaking to new officers at mess night at the Basic School, he begins by saying he is retired, so of course he has no influence on the Corps. It is clearly a joke. What bothers him most about today's military, he goes on to say, is careerism. It has eroded the other services, and it is creeping into the Corps. The only thing you should worry about, he tells the assembled second lieutenants, is taking care of your people. In fact, he recommends adding a new little box to the officer evaluation reports: It would say, Does this officer care more about his career than about his troops? A 'yes' mark would terminate that officer's career.
>
> *Tom Ricks*
> *"Making the Corps" (pg. 147)*

Unless you care more about others than yourself, you will fall prey to careerism

General Gray differentiates between those individuals who weigh the impact every assignment has on their career with those who accept whatever they are asked to do, placing their trust in the organization and its leaders. He took every assignment he was given and tried to make the most out if it. Helping others get ahead was more important to him than his own career.

In an interview, he commented on the difference, noting that people who focus on their careers may be as successful as those who place the priority on their people then added "but when they get to my age, I don't think they will be as happy." (He had a great smile on his face when he said it).

General Gray transformed the Marine Corps by his unique leadership, but his "greatest legacy" and the source of his happiness is the people he helped develop along the way, many who followed in his footsteps (as Commandants, Generals, and leaders at all levels). Even today (over 20 years after he retired as Commandant) there are still leaders in the Corps who consider him their mentor and role model.

As General Gray says, "both types of leaders can be successful." But, as evidenced by the many notes, birthday and holiday wishes he still receives (20 plus years later) from people at all levels *caring more about others than yourself* might just make you happier.

We expected to retire as a lieutenant colonel. Jan, who I was dating at the time, wouldn't have had anything to do with me if she thought I'd get promoted. The first change in that system happened when they deep selected me for colonel.

Then it was going to stop at colonel. Marine Corps Headquarters gave me the opportunity to go to Okinawa unaccompanied for a year and then come back and go anywhere for three years until I retired. 'That way you can take care of your mom.' (General Gray had been taking care of his mother ever since his dad passed away in 1968.)

And I said, 'You got it.' That was the deal. (Any time you get a chance to serve with the operating forces, you should jump on it.) So that was the plan. So instead of going to the Pentagon or some other career-enhancing assignment around the flagpole, I was going to go to WESTPAC, do what operational Marines do, come back, serve two or three years somewhere; wherever they had a billet, and then retire. There was no other plan.

General Al Gray

Never say "I" unless you are talking about a mistake

It is amazing how the General consistently gives credit to others in all the accomplishments (saying "we"), while taking responsibility for a mistake or an error in judgment made (saying "I").

> As you go along the trail, it's always useful
> to try to convince people that it's their idea.
> It really helps. It's a "we" kind of a thing.
> And that helps you too. The only time you
> want to say "I" is when there's a mistake.
>
> *General Al Gray*

He had a disdain for those who used what he calls the *vertical pronoun*, "I," instead of "we."

As a result, this Grayism has two corollaries…

You can accomplish anything you want if you don't care who gets the credit

General Gray calls it "mentoring development," operating behind the scenes in order to share the credit, adding, "nothing gets in the way of a good idea more than people who want to take the credit."

No one makes a mistake on purpose

We tend to over-supervise people and become concerned about how our effort will look when they make mistakes. We must turn our people loose and let them take actions. That is how they learn.

Mistakes are to be minimized, but not eliminated (that's a "zero-risk" mentality). People can learn from mistakes. And learning from mistakes is the best way to reduce future mistakes. Accepting mistakes can be seen as an investment in your people.

The greatest mistake a person could make can be the unwillingness to move forward because it involves risk. The failure to act (a mistake) leads to mediocrity. And in a world moving forward at a rapid pace, mediocrity leads to failure.

Should someone ever make an "intentional" mistake? No, as General Gray says: "no one makes a mistake on purpose."

If you are not teaching, you are not leading

General Gray says: "those who can't teach – can't lead."

The greatest lessons are passed on via "foot locker education." The key is in small unit leaders teaching their people. There is no substitute for the tested veterans telling how they survived and how things work. General Gray equates teaching to leadership and he set the example. The number of Marines, at all levels, that consider General Gray their teacher and mentor is legendary.

And he considered many of the legends of the Marine Corps his mentors. "I was privileged to serve and to know all the Commandants since enlisting in 1950."

And with teaching comes training – and another Grayism...

You can never spend too much money on training

Some say, especially in tough economic times, they cannot afford the costs of training their people. They seem to think training can be postponed until things get better. But General Gray says – "Training and education are the essence of preparedness."

When asked by Congress about putting money into the "quality of life" projects instead of training, General Gray responded that "…the best quality of life, was having a life," so he continued to put his "money into troop training."

The next level of authority is your natural enemy

General Gray believed the person best able to make a decision is the one closest to the situation, especially when the next level of authority may be miles away, or even in another country. Gaining approvals from higher levels can greatly reduce momentum and cause needless delays.

While this Grayism may be more "tongue-in-cheek" than the others, it clearly shows General Gray's dislike for what can be overly bureaucratic structures. "Systems can aid and abet mediocrity. Bureaucracy, while needed," he would add, "should be minimized whenever possible."

With a smile General Gray says, "Headquarters has an insatiable appetite for information. Give them what they ask for, but pay no 'gratuities'." When General Gray was Commandant and became "higher headquarters," he followed his Grayism – "cutting 10% per year (staff and resources) out of headquarters and sending them to the field."

Dave Reist recalls this story...

> In the late 1980s our Marine Expeditionary Unit (Special Operations Capable) [MEU (SOC)] was operating in France. General Gray was the Commandant and came to visit. Dinner at a local restaurant with a head table was arranged.
>
> When General Gray arrived with a small contingent (that included Brigadier General Jenkins) the MEU (SOC) Commander (a Colonel) attempted to get General Gray to move towards – and sit at – the head table.
>
> General Gray wanted nothing to do with it and eventually yelled across the room to Brigadier General Jenkins 'Harry, you're a General, aren't you?' Not knowing what was coming next, Brigadier General Jenkins responded 'Yes Sir.'
>
> General Gray said 'They have a seat at the head table for you' – Brigadier General Jenkins proceeded to the head table and General Gray sat with the lieutenants.
>
> *Brigadier General Dave Reist*
> *USMC (Retired)*
> *Senior Fellow*
> *Potomac Institute for Policy Studies*

Good staff leadership makes a difference

You know, there's nothing better than going and sitting in a commander's office or talking to him out in the field and kind of Marine-to-Marine saying, 'What are some of the problems? How could we help you?' It's really not a new idea. It's called "staff leadership." But it quite often is neglected because staff are so busy.

Colonel Al Gray
G-3, 2nd Marine Division
January, 1973

Do we really need to do this?

General Gray is a constant innovator. He is, as he signed a recent email, PM – in "perpetual motion." Serving in the Marine Corps for 41 years he, and those he led, never became complacent. He constantly asks – "Do we really need to do this? Is there a better way to do it? What prevents us from doing it?"

He is known for asking these three questions in meetings. He did in the Marine Corps and he continues to ask them today as he serves on various organizational boards.

> ...you treat everybody the way they ought to be treated in this business of helping other people, whether they are senior or junior. Over time, if you're consistent enough, people figure out you're really talking about what's best for the Marine Corps and not what you want to do.
>
> And a lot of it works out, so you can do these bold things. But you want to be right. You may not always get approval for what you want to do, for whatever reason, but you need to be right.
>
> *General Al Gray*

And along with being right, is taking risks – another Grayism...

There is nothing worth seeking that doesn't involve risk

General Gray believes that a risk-averse (or zero-defect) mentality is the antithesis of leadership. It stifles innovation, motivation, and leadership – "Chaos and uncertainty create opportunities."

Consider Certainty/Uncertainty as a continuum. Are you seeking Certainty (low risks), or are you comfortable with Uncertainty (without creating certainty)?

Certainty is described as: following established policies and procedures; making consistent and safe decisions (low risk); minimizing mistakes; thoroughly planning before acting; and building on *what is*.

Uncertainty is described as: creating new paths; making bold decisions; learning from mistakes; acting quickly (exploiting opportunities); and building on *what can be*.

Intent is the glue that holds everything together

Every mission has multiple parts: the task to be accomplished (the what) and the intent (the reason why), along with the desired result (of the action or intent). Of the two, the intent is predominant. The intent must convey the leader's vision.

It must be clearly understood at least two echelons or levels above and two echelons or levels below.

This permits subordinates to take actions quickly and decisively when the opportunity arises and when it is consistent with the overall intent.

> Marines will do what ever you ask them to, if you tell them why – unless intent is known, you are playing with half a hand.
>
> People know what they like doing, but they don't like being told how – telling them how restricts creativity. In every new assignment, I would talk to all the officers and staff NCOs about my thought process. It is important that intent is known.
>
> *General Al Gray*

Knowing intent allows the individual closest to the situation to determine how a task can be accomplished. Intent is a prerequisite to decentralized execution. Knowing the intent – the why, the how doesn't matter.

> The burden of understanding falls on the senior and subordinate alike. The senior must make perfectly clear the result he expects, but in such a way that does not inhibit initiative. Subordinates must have a clear understanding of what their commander is thinking. Further, they should understand the intent of the commander two levels up.
> *FMFM 1 "Warfighting"*
> *USMC Philosophy*

You communicate by how, as much as what, you say

> We believe that implicit communications – to communicate through mutual understanding, using a minimum of key, well-understood phrases or even anticipating each other's thoughts is a faster, more effective way to communicate than through the use of detailed, explicit instructions.
>
> *FMFM 1 "Warfighting"*

FMFM 1 explains General Gray's way of communicating, implicitly. First by establishing long-term working relationships to develop the needed familiarity and trust. Key people should talk directly to one another, whenever possible. How we talk (our tone and inflections) impacts our message. A leader should lead from well forward to gain an intuitive appreciation for the situation.

> When you lead from the front, you never have to ask 'what's going on?'
>
> *General Al Gray*

Communications without intelligence is noise... Intelligence without communications is irrelevant

> *Al Gray*

Master Sergeant J.D. Baker recalls the following story:

> In 1989, Marines were conducting combat operations in Panama. We were responsible for running combat patrols in and around the Arraijan Tank Farm, located on the Pacific side of the Canal Zone. Marines operated out of a fire base located in the tank farm. Few if any high–ranking officials ever made an appearance at the fire base – except one.
>
> As we were preparing for night operations, in came a couple of vehicles. We heard the CMC was coming to visit, but we never expected to see him at the fire base. Obviously, we were mistaken. General Gray had been at the Naval base looking to see all the Marines. When he found that most were in the fire base, he came out to be with the Marines.
>
> General Gray's presence lifted the spirits of all the Marines. He did not just hang out in the Combat Operations Center. He made his way out to all the fighting positions and spoke with the Marines going to conduct night ops. He personally thanked the Marines for the job they were doing and said if they needed anything to send it up the chain.

General Gray's visit spoke volumes for the Marines! They felt they had become comrades with the CMC because he took the time to meet and speak with everyone in the fire base. By the looks of his staff members, this was not part of the planned itinerary. The presence of General Gray was felt by all!

Master Sergeant J.D. Baker
USMC (Retired)

What are you interested in?

Earlier in his career, the future General Gray was one of many who had to brief a senior officer. Everyone else felt limited by the 10 minutes they were allocated to tell the senior officer everything that was going on in their area of responsibility. When it came to the future General Gray, however, he began his report by asking the senior officer – "What are you interested in?"

When Colonel Bob Nichols (later a Marine Lieutenant General) was assigned to be the Chief of Staff at Quantico in 1969, he received the usual briefings by the various departments and divisions of the Quantico commands. Each group presented their programs and projects. When it was his turn, Lieutenant Colonel Gray invited Nichols to have a seat and asked the Colonel if there was anything he wanted to discuss. The Colonel's schedule for the rest of the day was changed and, several hours later, the Colonel departed after having discussed a variety of issues and challenges.

General Gray tells this story:

> In 1976, Newt Gingrich was a fairly new Congressman and very interested in military reform and maneuver warfare. He asked the Pentagon if they could have a flag or general officer brief him. He wanted to talk to them and ask questions. They didn't really want to touch it.
>
> They didn't want to get involved with it. So they kicked it around here, there, and everywhere, and finally it ended up in the

Marine Corps, and the Commandant asked me if I would do it. Because he knew that I could do this kind of thing, maybe, without tying the Marine Corps to it as a service. In other words, it would be my own view and I could protect the Marine Corps and protect our principles and what we're all about.

You know, when you give 8,000 briefings to NATO and everybody else, you ought to be able to do that. In those days, I had the 4th Marine Amphibious Brigade (MAB) – our Atlantic and Mediterranean contingency force – home ported in Norfolk.

I drove up to Arlington, Virginia on a Sunday. Gingrich was living in an apartment in Arlington, and he and his wife were there. We went from 9:30 in the morning until 1:30 the next morning. He asked me all kinds of questions and I told him what we thought and he asked about reform and movement and maneuver warfare, and on and on and on.

Of course, he never forgot that. He remembered it and later he came down to Camp Lejeune and visited the 2nd Division while I was down there and stayed out at the beach house. So that was a good relationship, and it's still good today.

General Al Gray

What you did isn't as important as what you were thinking

Critiques are an important part of Marine leadership. A key attribute of leaders is their ability to review their decisions with clear reasoning. Critiques provide the reasoning.

Critiques should focus on the rationale for what was done (not just what was done) – what factors were considered, or not considered, in making an estimate of the situation.

General Gray encouraged Marines to "take their rank off" to freely discuss the situation and what they were thinking. Mistakes are essential to the learning process and should be "cast in a positive light."

Watching General Gray closely, he is a rare leader who critiques "on the go," processing information as he speaks.

Understanding Tactics

> In tactics, the most important thing
> is not whether you go left or right,
> but why you go left or right.
>
> *Major General Al Gray*
> *CG 2nd Marine Division*
> *FMFM 1-3 "Tactics"*

Leadership by walking around

His greatest source of information comes from another Grayism: *Leadership by walking around*. Marines called him "the Gray Ghost," often showing up at unexpected places and times.

> In the fall of 1977, 4th MAB had just begun a series of operations in Northern Europe. Before the exercise was to commence, the German Guards sounded the alarm that an intruder was attempting to enter the compound, the EW Officer rushed to the sentry position. There was General Gray in full tactical gear. The sentry was adamant that no one was entering during his watch. After some translation and intervention by the German Unit Officer, General Gray went to see his Marines.
>
> The General proceeded to each position to be briefed by the Marines, who commented on the exchange with their NATO counterparts. Each Marine briefed on his equipment and any shortfalls, or problems. The General then went to each of the German counterparts and debriefed them. His last statement was 'I want a message outlining each piece of equipment, its problems/shortfalls and recommended fixes.'

What did I learn from General Gray? Leadership by walking around, especially unannounced, talking to all the Marines, finding out what they know, what they have been told and how much latitude they have to do their job, will tell you volumes about the small unit leaders under you. And when stopping to talk to a Marine, talk to him like there is no one else there and let him know that you are very interested in what he is doing and why.

Colonel John Bicknas
USMC (Retired)

Do not make any more enemies than you already have

Colonel Tom O'Leary, USMC (Retired) says his favorite Grayism is what he called – the Conservation of Enemies. Here's how Colonel O'Leary tells it:

> When General Gray was Commanding General of the 2nd Marine Division, I remember him telling us about the principle of Conservation of Enemies. He said first do no harm to the people and try to go to bed with fewer, or at least no more enemies, than you had when you woke up. This was good advice for both combat and staff.
>
> *Colonel Tom O'Leary*
> *USMC (Retired)*
> *Executive Vice President*
> *Potomac Institute for Policy Studies*

In Vietnam, 1967, Major Al Gray said:

> There are two fundamentals when you're in a guerilla and counterinsurgency warfare environment. One of them is, you never do anything that's not good for the people you're trying to help. And number two, you don't ever make any more enemies than you've already got.
>
> *Major Al Gray*

You can move elephants under their own power

The statement would be great advice (and insight) from any leader. It's especially "powerful" when you realize the statement comes from the man, who as the 29th Commandant, was in command of nearly 260,000 Marines (Active and Reserve).

Managers may rely on the power of their positions. Leaders realize that power used is power lost. Some call it being empowered (power given). We prefer power assumed. Whatever you call it, people moving on their own power can move organizations, small and large (as well as elephants).

General Gray says:

> Everyone moves on their own power not yours. And the other thing that I always fervently believed in and I've said it 2,000 times in speeches – I've never met a Marine, officer or enlisted, a good one, that couldn't do 400 percent more if we let them.
>
> *General Al Gray*

Which leads to a series of related Grayisms:

Turn people loose

– When asked about how he transformed the Marine Corps, General Gray says he "…provided the guidance and then just turned the Marines loose. They did the rest."

Over-supervision causes more problems

– People can't grow – can't think, if over-supervised.

Develop the best in everybody

– It's harder to lead people with great potential.

Achievement feeds on itself

– And creates more achievement.

You can lead the Corps like you do a platoon

— More people requires more trust.

To which General Gray would add:

> It would be negligence if I didn't know how to get things done. If you don't know something, you find out, and then you make it happen.
>
> It's a gift, to have a lot of good people around you. So, I was always throwing stuff at them, to make them go out and do. And that's how you get a lot more done.
>
> *General Al Gray*

Additional Grayisms

These additional Grayisms are simple statements, worth repeating, that reinforce other Grayisms and require less explanation.

Some may be quick to point out that they have heard the same, or similar, statements from other leaders. General Gray would tell us – "who gets the original credit is not important."

The tribute to General Gray's leadership is the frequency that these statements, like the ones already discussed, are attributed to him, remembered, repeated, and shared with others.

Go with your gut feelings

– Follow your instincts.

Everyone kicks one in the grandstands once in awhile

– Turn people loose. People who do a lot will make mistakes.

Mental toughness is more important than physical toughness

— It's mental toughness that gets you over the hurdle when you are exhausted.

Marines do not have a corner on ideas

— Take ideas from any source.

You must be ready to go tonight with what you have in your pocket

— It's being expeditionary.

People are like kids at a ballgame

— They can see through those who are only going through the motions.

Treat people the way you would like to be treated

– The Golden Rule.

Take all the help you can get

– There are no crowded battlefields.

Do your homework

– Be prepared.

Wherever two Marines are, one is in charge

– And one is following.

Spend a little time every day making sure you have people who could replace you

– Train and educate your own replacement.

It's not just how quick you can get there, it's how quick you can do something when you get there

– Be light enough to get there and heavy enough to win.

It's a poor carpenter that doesn't hit his thumb once in awhile... he isn't driving many nails

– The very best make mistakes. If you don't push yourself to your limits and beyond, you will limit your true potential.

Mistakes happen

– The key is to learn from them.

Fix the problem first

– Then fix the process later.

"Institutionalize" ideas

– Until a better one comes along.

Trains go down the tracks

– You can get on, or get off.

Failure is an orphan

– While success has many followers.

Loyalty goes down

– As well as up.

You lead people

– And you manage assets.

A good plan well-executed is better than a perfect plan delayed

– Once things start happening, the plan goes out the window. Be ready to adapt.

Whenever you make assumptions, you need an alternate plan

— You can't predict the future, you can predict trends.

If you have any question about which alternative to choose

— Pick the one I told you to do.

You invest in an idea by the people you place in charge

— They are the best investment you can make.

People are like weapons

— They have capabilities and limitations — you have to know them both.

Never ask people to fly beyond their own comfort zone

Don't be afraid to learn from others

I don't want a conversation. I just want an answer

You shouldn't kid people unless you really like them

It's the art and science of getting things done

– There are no textbook answers.

And the most lasting Grayism

**Do as much good as you can,
for as many people as you can,
for as long as you can.**

And a special Grayism – For his very special wife, Jan Gray:

When asked how he has remained married for so many years, General Gray said (tongue in cheek):

…you have to choose…

You can be right, or you can be happy.

To which Jan Gray would add:

P.S. Be careful what you ask for.

Obviously, General Gray has made the right choices.

Additional Stories

As noted at the beginning of this book, we have captured only a small part of General Gray. Ask any Marine that served with him or who have heard stories about the General, and odds are they will have a story, a Grayism, to share with you.

And, lastly – Grayisms and the accompanying stories could be told in many different ways, in different order, and reinforced with different stories. There are many, many more stories, including the three additional ones that follow.

There are surely additional Grayisms that I have failed to capture. And if a Grayism, or story is different than you remember it, I am responsible for the compilation. Please do not let any of my mistakes distract from your enjoyment of reading about and hopefully gaining greater insight into this very special Marine – General Al Gray.

The following three stories reinforce more than one Grayism…

From Lieutenant Colonel Eric Carlson:

> Early one morning back in 1983 when General Gray was the Second Marine Division Commander, he climbed up the fire escape ladder into the Second Battalion Sixth Marines operations shop and surprised us all! Over the three decades that have passed since that encounter, the General has often taken unexpected leadership actions that endear him to those surprised by his presence.
>
> One often overlooked aspect of General Gray's character is his innate ability to love his neighbor and quickly build enduring trusting relationships of significance – be it with a President or a Private or an elderly neighbor in Southeast Washington D.C.
>
> During our oral history interviews, the General shared some remembrances of his time as Commandant of the Marine Corps. Shortly after moving into the Commandant's House at Marine Barracks Washington, General Gray was doing some yard work in front of the house one Sunday afternoon. He saw an elderly lady stop to read the building's historical marker facing the street and then just stare at the building...

'So I went up to her, and I said, 'Ma'am, are you okay?' And she says, 'Oh, yes.' I said, 'would you like to come in and see the house? They are not here this weekend and I would be glad to give you a tour of the house.' She said, 'Oh, could I?' So, with her thinking I was the gardner, I took her in and gave her a tour. When we went back outside, she started crying.

I said, 'Ma'am, are you all right?' She says, 'Oh, yes. You don't understand. I've walked by this house every day since I was a little girl and I've dreamed about what was in the house. And now I've seen it.' I asked, 'Where do you live?' She said, 'I live over on Ninth Street, about a block and a half from here.'

So Monday we sent some Marines out, down four blocks on each side of the Commandant's house to invite everybody to a parade the next Friday night. Security and everybody went a little off here, and they were really worried. But I said, 'No. Don't worry about it.'

[Everybody had a good time, and after that the Commandant could go anywhere he wanted in southeast Washington. There wasn't anyone going to bother him.] That's where I used to walk and run with the dogs at 4 o'clock in the morning.'

•

While that story touched our hearts, the strategic significance of trust relationships with ordinary citizens is also often overlooked. When asked next about his affinity for 7-Eleven coffee (and frequent ventures out to the local 7-Eleven while Commandant), the General continued…

'I used to always like to be living somewhat near a 7-Eleven store because I like to go there and see what the people are talking about and have my coffee, and I've always done that – even now. I remember when we were having the strategic arms reduction talks with the Chairman of the Joint Chiefs of Staff, the Joint Staff, the President, and much of the Cabinet were all in the room. We were briefed on how we were going to be reducing and destroying nuclear weapons.

Then the next briefing came along and said how we were going to get new nuclear weapons. I remember saying – when it got around to me, 'I understand these briefings, but I'm not sure that the American people would. I don't think they'd understand this down at 7-Eleven.'

And Admiral Crowe said to me, 'What the hell are you talking about?' And the President was looking. I said, 'Well, I go to 7-Eleven over the weekends and get a cup of coffee and I talk to the people there. You know, they understand these strategic talks a little bit, but they're really more interested in why they can't get a new county road built, and why they pay more FICA than they do income tax. And I said, 'That's the kind of thing they're interested in.'

Then I added, 'They're going to have a little trouble understanding why we're going to destroy nuclear weapons – spend money to destroy nuclear weapons – and then spend more money to buy new nuclear weapons. So we need to figure out a way to articulate this a little better.' That got the message across.

Lieutenant Colonel Eric Carlson
USMC (Retired)

From Scott Laidig:

> Several of us have had the privilege of attending the annual Wallow of the Military Order of the Carabao as General Gray's guests. It is a formal event attended by about 1,500, including many active and retired general and flag officers and other dignitaries.
>
> The Marine Band provides the musical entertainment. After the National Anthem, a medley of service songs is played. It's not surprising, that members from each service rise and either stand at attention or sing along with their song. At the first Wallow we attended (early 1990s), we were seated as the Band played the Army's "Caissons Go Rolling Along."
>
> We were surprised to see General Gray standing at attention! Following his example, we all immediately stood. However, we saw no other table standing for all the service songs.
>
> Fast-forward nearly 20 years: there are many, many (and nearly all the Marines) in attendance, now standing for the entire medley. General Gray's example, without a single spoken word, has become the norm rather than the exception.
>
> *Scott Laidig*
> *USMC (Retired)*
> *author of "Al Gray, Marine"*

From Sergeant Robert Jones (as told by Scott Laidig):

> Rob Jones was still a Corporal and had arrived at Bethesda only a day or two prior to General Gray's first visit to him. Rob was a recent graduate of Virginia Tech; in 2010, he had been serving in Afghanistan as a combat engineer attached to 3rd Battalion, 7th Marines. While he was investigating an Improvised Explosive Device (IED), it exploded, leaving Rob a double amputee.
>
> But Rob has never seen an unhappy sunrise and he lifts all around him with his optimism and good cheer. There in his hospital room at Bethesda, Rob was talking quietly with his dad, when an unknown but pleasant looking older gentleman in coat and tie walked in.
>
> 'Hi, I'm Al Gray,' the newcomer announced.
>
> Rob recalled thinking, 'Do I know this guy?'
>
> Some pleasant chitchat followed while Al Gray tried to find out if everything was okay with Corporal Jones, if he needed anything, and if the doctors were taking good care of him. Finding nothing amiss, "Mr. Gray" shook hands with Rob's father and Rob, and then handed the wounded Warrior a commemorative coin that denoted the 29th Commandant of the Marine Corps.

Rob studied the coin, and then asked, 'How long have you had this?'

'Oh, probably 20 years or so,' replied Al Gray.

'Did you really get it from the Commandant,' asked a disbelieving Rob Jones.

'Oh, that's me,' chuckled Al Gray in response as he left.

And in that way, Sergeant Robert R. Jones, United States Marine Corps, joined a mass of others who might say, 'Who doesn't know General Al Gray?'

Al Gray, Marine, forever continues to do as much as he can, for as many as he can, for as long as he can. They really are the words he lives by.

Sergeant Robert Jones
USMC
(as told by Scott Laidig)

General Al Gray Biography

In 1991, General Al Gray retired from the U.S. Marine Corps after 41 years of service. From 1987-1991, General Gray served as a member of the Joint Chiefs of Staff, was the 29th Commandant of the Marine Corps, and was advisor to both Presidents Reagan and George H.W. Bush. As Commandant, he institutionalized and published a Warfighting Philosophy for the Marines. General Gray developed and implemented a new long-range strategic planning process for the Marine Corps, established the Marine Corps University, and implemented other longstanding changes, such as ensuring that every Marine is a rifleman first and the Marine Corps was special operations capable.

General Gray enlisted in the Marine Corps in 1950 and achieved the rank of Sergeant while serving in amphibious reconnaissance with the Fleet Marine Force, Pacific, aboard the submarine USS Perch (ASSP-313). He was commissioned a Second Lieutenant in 1952.

Service in Korea included commands of both artillery and infantry units. In 1955, he commanded the 4.2 Mortar Company in the 2nd Marine Division at Camp Lejeune, NC. From

1956-1961, he served overseas in Japan and throughout the Far East in special intelligence command activities. During this period, he pioneered the Marine Signal Intelligence/Electronic Warfare capability for Marine Expeditionary Forces. He served at Guantanamo Bay during the Cuban Missile Crisis of 1962 and commanded the first Marine ground unit to operate independently in the Vietnam War in 1964. He returned to South Vietnam in 1965 and served through January of 1968. Assignments included S-3 of the 12th Regiment, Commanding Officer of the Composite Artillery Battalion and the U.S. Free World Forces at the Gio Linh Outpost along the Demilitarized Zone and Commanding Officer of the First Radio Battalion, Vietnam. After service with the Defense Special Communication Group in Washington and duty with the Development Center at Quantico, VA, he returned to Vietnam in conjunction with Intelligence, Surveillance, and Reconnaissance matters.

From 1971-1973, he served with the 2nd Marine Division as Commanding Officer of the 1st Battalion 2nd Marines where he led them through a Mediterranean deployment. He then served as the Commanding Officer of the 2nd Marine Regiment and subsequently, as the G-3 (operations, plans, and training) for the Division. After graduating from the Army War College in 1974, he returned to the Western Pacific where he commanded the 4th Marine Regiment and Camp Hansen in Okinawa. He had a primary role in the planning and execution of the Southeast Asia evacuation operations in 1975.

From 1976-1978, he commanded the 4th Marine Amphibious Brigade and, concurrently, the Landing Force Training Command, Atlantic, which included extensive operations on the northern and southern flanks of Europe in their role as strategic reserve for NATO. From 1978-1981 he was the Director of the

Research and Development Center at Quantico, Virginia. As Director, he was responsible for the development of concepts, doctrine, tactics, techniques and equipment employed by landing forces in amphibious operations. He was the primary driver in the development and acquisition of Light Armored Vehicles for the Marine Corps and the leading proponent for the development of the High Mobility Multipurpose Wheel Vehicle (HMMWV) for the armed forces. He streamlined and accelerated studies and doctrinal development. General Gray directed development of the first approved Mission Area Analysis for Amphibious Warfare.

During the timeframe of 1981 to 1984, General Gray was the Commanding General, 2nd Marine Division at Camp Lejeune, North Carolina. General Gray was responsible for the total performance of a large ground force trained and prepared to execute various types of missions. From 1984-1987, General Gray was the Commanding General, Fleet Marine Force Europe, II Marine Expeditionary Force and Marine Striking Forces Atlantic (NATO). In this capacity, General Gray exercised command of all Marine Corps operational forces. Marine Security and Ship Detachments assigned east of the Mississippi River extending to and including the Atlantic Ocean, Norwegian, Caribbean and Mediterranean Seas and Europe.

Since retirement from the Marines, General Gray has served on several corporate boards, both public and private, as an advisor or director. He is the past Board Chairman of three public companies and three private companies. He currently serves as a Senior Fellow and Chairman of the Board of Regents for the Potomac Institute for Policy Studies where he is also a member of the Board of Directors.

Consistent with his longstanding interest in education and public service, he serves as Chancellor of the Marine Military Academy, and as a mentor to the Defense Science Study Group. He is Trustee for the American Public University System and is Chairman Emeritus of the American Military University. He also is a Trustee Emeritus of Norwich University and a past Trustee of Monmouth University. In addition to past service on the Defense Science Board, he has been an advisor to the National Reconnaissance Office and a member of the National Security Agency Advisory Board since 1992. General Gray is Director of the Marine Corps-Law Enforcement Foundation. He is also Honorary Commandant of the Marine Corps League.

In addition to being a recipient of numerous American military awards, General Gray has also received awards from the Republic of Korea, the Netherlands, Chile, Argentina, Colombia, Brazil, and the Philippines.

General Gray holds a B.S. from the University of the State of New York. He also attended Lafayette College, the Marine Corps Command and Staff College, Army War College and did graduate work at Syracuse University. General Gray is the recipient of two honorary Doctor of Law degrees, one from Lafayette College and the other from Monmouth College, and is a recipient of a Doctor of Military Science degree from Norwich University. He was the first awardee of an Honorary Doctorate of Strategic Intelligence degree from the Defense Intelligence College (now the Joint Military Intelligence College). He also has an Honorary Doctorate from the Franklin University, and an Honorary Doctorate in public service from the American Public University System.

GENERAL AL GRAY AND THE SEMPER FI FUND

Over the past decade, General Gray has served as Chairman of the Semper Fi Fund, which helps our wounded Warriors and their families.

The Semper Fi Fund (a 501(c)(3) nonprofit) and its affiliate program America's Fund are set up to provide immediate financial assistance and lifetime support for injured and critically ill post-9/11 service members from all branches of the U.S. Armed Forces and their families.

> In 2004 when we incorporated the Injured Marines Semper Fi Fund, affectionately called the Semper Fi Fund, General Gray was the very first four Star General to believe in our efforts and roll up his sleeves to help.
>
> He, without hesitation, agreed to become our Chairman of the Board of Directors. For this we are forever grateful. We were an unknown group to so many, but General Gray believed in our mission to the depth of his soul and he believed in our team!!
>
> We are going on over 10 years of service to our ill and injured service members and their families with still an extraordinary amount of service yet to give. We are Blessed to have General Gray at the helm!
>
> *Karen Guenther*
> *President, CEO, and Founder*
> *Semper Fi Fund*

No American Warrior or their families must ever feel alone on the long road to recovery.

Al Gray, Marine

To U.S. Marines

When General Gray first spoke after assuming command of the United States Marine Corps on 30 June 1987, he spoke at length paying tribute to his predecessor and long-time personal friend, General P.X. Kelley. He then turned to the Marines in parade formation, representing the entire Marine Corps, and said:

> "This great nation loves her Corps of Marines; They pray for us; they support us; they fund us: they also place a couple of demands on us. The nation demands that you and I and all others like us be a little bit special. The nation demands that we be the best led, the best trained, the best disciplined, particularly self–disciplined, force on earth. The nation demands that we teach nothing but winning in battle and in life. We can do that tonight. And by God, you are going to make it happen in the years to come. God bless and Semper Fidelis."

General Al Gray
Commandant of the Marine Corps
30 June 1987

Additional Information

Grayims have been a source of "Leadership Minutes" and used in leadership development modules and in other materials and writings by Paul Otte and published by the Ross Leadership Institute. To subscribe to daily Leadership Minutes and learn more, visit: www. rossleadership.com.

Paul Otte and the publisher would like to thank the following individuals for their contributions to this work: Mike Swetnam, Robert Jones, Eric Carlson, Scott Laidig, Dave Reist, Tom O'Leary, John Bicknas, JD Baker, John Sheehan, George Flynn, David Sommers, and Sherry Loveless.

Additional books by and about General Gray are available:

General Al Gray and Paul Otte. *The Conflicted Leader and Vantage Leadership.* Franklin Leadership Press: Columbus; 2006.

Scott Laidig. *Al Gray, Marine: The Early Years 1950–1967, Vol. 1.* Potomac Institute Press: Arlington; 2013.

Please contact The Potomac Institute for Policy Studies for more information- Email: webmaster@potomacinstitute.org.

Works Cited

FMFM 1 Warfighting. USMC; 1989. PCN 139 000050 00. www.theusMarines.com/downloads/FMFM1/FMFM1-1.pdf.

FMFM 1-1 Campaigning. USMC; 1990. PCN 139 000060 00 www.theusMarines.com/downloads/FMFM1_1/FMFM1_1-1.pdf.

FMFM 1-3 Tactics; 1991. PCN 139 000104 00.

John C. Chapin. *Uncommon Men, The Sergeant Majors of the Marine Corps.* White Mane Books: Shippensburg; 2007.

Thomas E. Ricks. *Making the Corps: 10th Anniversary Edition with a New Afterword by the Author.* Scribner: New York; 2007.